Here's what reviewe...

Insightful resource laced v... ... ...
inspiration for all to embar... ...p... ...c.. own unrealized journeys.

– Stephanie Vay
NASA Research Scientist and Native American Indian Dancer

Ann is a remarkable woman to have taken on and conquered a mountain. Her kindness speaks volumes as she loans her equipment to another climber suffering from the cold. Ann has completed the journey of a lifetime.

Katherine Read
Lynnhaven Dive Center
Virginia Beach, Virginia

We have a tendency to forget that mind, body and spirit faithfully wait for us to challenge ourselves whether age eight or eighty. The courage to act, then, is the obligation we all have to grow and experience life in its ever infinite glory and wonder. Taking any new first step in that challenge is life ... is the journey. Ann shows us the way! Thank you.

Pete Cawley
Ex Lt. Navy Fighter Pilot
Flight Officer United Airlines

A Trek to the Top of Mount Kilimanjaro

Africa's Highest Mountain

by

Ann Brand, Ed. D.

Virginia Beach, Virginia

Copyright © 2003 Ann Brand

Photos by Ann Brand unless otherwise indicated

Published by:
ECO IMAGES
P.O. Box 61413
Virginia Beach, Virginia 23466-1413
Email: wildfood@infionline.net
URL: http://wildfood.home.infionline.net

ISBN: 0-938423-10-X

Available from:
Ann Brand
2107 Wake Forest St.
Virginia Beach, VA 23451

All rights reserved, including the right of reproduction in whole or in part in any form or by any means, electronic or mechanical.

Printed in the United States of America

First Printing

IN APPRECIATION . . .

To my editors, Betty Ann Chandler, Alysia Hedrick Grimes and Vickie Shufer, for without their time and encouragement this book would never have happened, and to Paul Shufer for his graphic arts skills and advice. Also thanks to Lynda Blanks, a friend at every turn.

I would also like to acknowledge the following for reading and critiquing the manuscript:

Lillie Gilbert, *Wild River Outfitters*
Katherine Read, *Lynnhaven Dive Center*
Judith Dockery, *Attorney at Law*
Robert Taylor, *System Analysis*
Pete Cawley, *United Airways*
Stephanie Vay, *NASA*
Dawn Bradshaw

Dedicated to my hero:

My mother, Dorothy Nicholson

Contents

Prologue ... ix

Preface ... x

Part One *Preparing for the Climb*

1 Invitation to a Journey ... 15
2 The Decision Making Process ... 19
3 First Meeting ... 27
4 The Agenda for Travel & Trek ... 31
5 Acclimation ... 33
6 From Here to There ... 35

Part Two *The Journey*

7 The Big Bird ... 43
8 First Glimpse ... 45
9 The Rain Forest at 9,450' ... 47
10 Machame Hut 12,464' ... 53
11 Shira Hut 12,595' ... 57
12 From Baranco to Barufu 14,924' ... 65
13 At Barufu Hut 14924' ... 69
14 What's Ahead At Midnight Barufu – 4,000' Ascent ... 75
15 Time to Turn Around ... 81
16 Home at Sea Level ... 85

Prologue

On July 8, when I returned to the Impala Hotel in Arusha, Tanzania, following the completion of my climb up Mount Kilimanjaro, I looked in the mirror at a dreadful sight. My swollen face and mouth were covered in blisters. My fingernails had a pearl-blue appearance and the backs of my hands were encrusted with boil-like sores. All of these came as no big surprise to me. I was not even concerned over the fact that I had been for six days without a shower.

What I had not anticipated was the devastating feeling that gripped my soul when I realized that we had to turn around with only a few hundred feet to go to the top. Although I knew that turning back was the right thing to do, it did not ease the disappointment.

Writing this book from my diary entries has provided me with the opportunity to share with you the provocative moments of the trek. It has given me a chance to describe biomes with their beauty and their bare devastation.

I do not declare it to be 100% factual for there were many times when I experienced lassitude and listlessness. A lack of oxygen plays tricks on your basic ability to logically reason.

Preface

A person said to me when I shared with him that I was training for a climb of Mount Kilimanjaro, "It is only a walk in the park." No Bob, it was not a walk in the park. Climbing anything from a mountain or set of steps is no "walk in the park." It requires that you have prepared your mind, body and soul for a journey that may bring enormous demands.

For anyone who reads my book, I hope that they will take it as an invitation to plan a "journey" that they have dreamed of making. If financial resources, family obligations, age, poor health or even a physical impairment are perceived to be a limiting factor, believe me, they are only part of the challenge. An obstacle can often turn out to be part of the solution. As long as one has books to read, music to listen to, and the desire to love and to listen to others a journey is well within the realm of reality.

When journeying, always become knowledgeable about the culture and the people you are visiting. Listen to their "body language," smile, and say "thank you" when they give you assistance. Learn how to say "thank you" in their language.

Lastly, never allow yourself to equate your individual self-worth with whether you finish your journey or not. Crossing over the line that one has designated as a destination, or returning with a gold ring is no measure of success or of failure.

Enjoy and find peace with your soul, one day at a time, as you move about in another's culture and environment.

Ann H. Brand, Ed. D.

Part One

Preparing for the Climb

"... one of them enveloped in a prominent white cloud ...
and it was as good as certain to me
that it could be nothing else other than snow."
– *Johann Rebmann, 1848*

1

Invitation to a Journey

Have you ever thought about doing something wild, wonderful, and wooly? Well, I did! No, I wasn't fresh out of college or graduate school when I gave the above some serious consideration: I was wrapping up thirty years in public education.

My phone rang early one morning and out came an invitation that would encompass "wild" and "wooly" but proved to be short of "wonderful." The male voice on the other end said, "Would you like to trek Mt. Kilimanjaro in Africa with four other professionals a year from today?" What a stunning proposal. I asked for one day to think about it. The actual climb was scheduled for July 3, 1999 from Arusha, Tanzania.

Mt. Kilimanjaro is Africa's highest point and consists of a cluster of mountains rising up from the plain to a peak at 19,340 feet. Located only 2° south of the equator on the northern boundary of

Tanzania, its snow covered peaks intrigued early explorers who thought snow on the equator was impossible. Snow-capped mountains, active volcanoes, lush rainforests, arid deserts, glaciers and cliffs, all contribute to the wealth of ecosystems found in Tanzania.

There are six proposed routes for reaching Uhuru Peak which is the summit of the mountain at 19,340'. They are *Marangu*, *Mweka*, *Umbwe*, *Machame*, *Shira*, and the *Lemosho* Routes.

The *Marangu* is the most popular route and takes five days and four nights. It is called the "coca-cola" or the "tourist route." It is described as a straight path with three good wooden chalets 'huts' along the way. The first two huts are manned by a permanent staff. The first hut has 60 beds, water, and plenty of firewood. The second hut has 120 beds, water, and some firewood. The third, Kibo, has 60 beds, no water, and no firewood. The trail to date has no difficulties. The last day of the climb for this route is stated to be "a little bit strenuous."

The *Mweka* Route is a second option; it takes 4 days and 3 nights. It is classified as being the most strenuous, direct and fastest way to the summit. Permits are obtained at Marangu Headquarters before you are dropped off at the College of Wildlife, near Mweka Camp. To reach the summit, you stop the first night at the Mweka Camp which has plenty of water and wood, the second at Barafu Hut with no wood but water, and the third day you climb to the top.

The *Shira* Route is the third possibility. It requires 5 days or more. You drive to Londorossi Glades Park Gate, get a permit to continue on, drive up to the Morum Barrier Gate in a 4x4 Rover and from there trek to Shira. From Shira you take the Machame Route for the duration of the trek.

The fourth choice is the *Umbwe* Route. It is classified as being relatively short and scenic; it is for 5 days and 4 nights. But it is not an easy ascent, it is quite steep almost all the way up. This route is recommended for only the very fit!

The *Lemosho* Route is the fifth option. It is slated for 5 days or more. After getting the permit at Londorossi Glades Park Gate, you drive back to Lemosho's starting point. The first day is very short, less than four hours. The stopping point is Shira Camp One. The next day you trek to Shira Camp Two for the night. From this point you proceed on the *Machame* Route.

The last choice is the *Machame* Route. It takes 5 or 6 days and is the route that we would take. However, we would make the trek in five days and four nights. It is described as a gradual ascent with a scenic view.

The trek was to revolve around a safari to Boswana which would commence following the climb. The schedule read as follows: from Norfolk, Virginia to Johannesburg, South Africa; spend the night in Johannesburg; the next morning, fly to Nairobi, Kenya; ride to Arusha, Tanzania by van; eat supper and sleep a few hours at the Arusha Hotel before leaving for the Machame Gate at daybreak to start the trek. From there we would trek through the rain forest to our first camp at Machame; from Machame through the damp forest zone with its trailing fronds of lianas hanging from the trees to Shira; from Shira to Barranco; from Barranco to Barafu; sleep a few hours, have tea, and trek 3,000' to watch the sunrise before climbing to the roof of Africa at 19,340'. The descent could be the *Mweka* or *Machame* Routes. We would descend the *Mweka* Route on the fifth day and continue on to the Horombo Hut where we would spend the night. The next morning we would hike to the gate. The sixth day to acclimate was removed from the schedule. I never asked why; I just trusted that the outfitters who were supplying our guides and cook knew what they were doing.

    The trek would require no technical climbing skills, but it would require physical and mental stamina, time, financial resources and a knowledge of all aspects of Altitude Illness. Acute Mountain Sickness (AMS) is the most common form and is characterized by nausea, headache, lack of appetite or sleeplessness in its mildest form. High Altitude Pulmonary Edema (HAPE) results from a lack of oxygen in the air combined with high pressure in the arteries, making breathing difficult. High Altitude Cerebral Edema (HACE) is thought to be caused by swelling in the brain as a result of oxygen starved vessels. Loss of balance or coordination are early signs of HACE, followed by a feeling of weakness. For me a touch of mountain mental madness would be a prerequisite.

2

The Decision Making Process

Lists, lists, and more lists…

Exactly how does one go about making a decision of this magnitude? Of course life is filled with decisions of one degree or another, but it's not often that one weighs the pros and cons of a trek of this proportion. My decision-making process consisted of several components: lists weighing benefits and costs, the counsel of a trusted friend, and a kayak trip to bring the friend and the lists together.

Being the obsessive compulsive individual I am, I started my decision process by making numerous lists, which I subdivided into more lists. The purpose of these lists was to give me a clear picture of the pros and cons for taking this trek.

After my lists were finished, I invited my best friend to join me for a kayaking trip. I always invite her to join me on occasions of personal importance. She loves to kayak and she listens attentively to every word I have to say. In fact, I have shared all my great and not so great moments with her. I have found kayaking to be an exciting way to make serious life decisions. My friend is not built for paddling, so this job is mine. Her stoic appearance attracts a lot of attention, which adds entertainment to any trip we take together and makes me forget that I'm doing all the physical labor. The last time we took a daylong outing, we kayaked to the "Narrows," four miles round trip, went ashore, and spent the day reading. I chose to read to her all one hundred and twelve pages of my doctoral dissertation.

The purpose of this kayaking trip was to share my endless lists with her and obtain an objective opinion of the entire idea. Besides seeking her opinion, my goal was to process my thoughts in an orderly fashion so that I could make a wise decision. "The paddling adventure will take an entire day," I explained to her. "I have a great decision to make." To my delight, she accepted my invitation without questioning me on the specifics of my decision.

Our kayak trip started at my front door. We rolled the kayak down the street on special wheels that facilitate the sand travel. The entry point is a small body of water called Long Creek. From the creek, we glided into the Lynnhaven and then out into the Chesapeake Bay. After getting into my paddling routine, I revealed to her the decision at hand, as well as my reasons for involving her in the process. "I've been asked to climb Mount Kilimanjaro in Tanzania, Africa," I said. "I have prepared lists of pros and cons that I need to sort through in order to make my decision. Decisions of this magnitude require contemplation, thought, and great consideration. I've always been able to depend on you as a tremendous listener."

It was time to get started with the lists. They were broken down into two main topics: "Why I Should Not Climb Mount Kilimanjaro," and "Why I Must Make the Trek." At this point I had written, rewritten, and reviewed my lists so many times that I no longer needed to look at them. That was fortunate, since the kayaking required my physical attention.

I decided to start with the longest and most logical list, which was entitled, "Why I Should Not Climb Mount Kilimanjaro." This list was lengthy and quite sobering. It contained legitimate, rational reasons why I should decline the invitation, ranging from financial to physical.

"Let's start with the financial costs involved in the trip," I began. "The plane tickets will be expensive. Then there is the cost of the actual trek, tips for porters, guides, and cooks, visas, hotel rooms in Johannesburg, South Africa and Arusha, and gifts for my family and friends. I'll need to purchase gaiters, capilene underwear, Gore-Tex rain hat, pants and jacket, special glacier glasses in my prescription, and adjustable hiking poles. I'll require shots and pills that are not covered on my insurance policy such as inoculations for polio, meningitis, hepatitis A and B, and cholera. I'll need Larium (to be taken prior to and following the trip), a 15 day supply of Cipro for any bacterial infections, and Diamox for altitude sickness," I noted. "Oh, and travel insurance to cover lost bags, canceled flights, and getting my body back to the states in case of a fatality. I'll definitely need a new camera with auto focus and a telephoto lens. No trip would ever be complete without a camera for those 'Kodak' moments," I added in my most convincing tone.

A Trek to the Top of Mount Kilimanjaro

"You see, there are those little items that are often of more importance than the bigger, more expensive ones," I stated. "Like, you know, ten pairs of socks (five cotton and five wool), two pairs of Bermuda shorts, five cotton T-shirts (two long sleeve), underwear, a ski jacket for the last day, a balaclava, gloves, hand warmers, a water bottle, energy bars, two flashlights, and a head lamp with batteries that will not freeze at sub-zero temperatures. That's all I can think of at the moment, although there may be other items that I'm not aware of yet," I added. Until hearing myself go through the extensive list of things I would need, I had not fully processed the financial implications.

Those were just the financial expenditures! There were certainly other costs, even more daunting, to consider. Up to this point, my friend elicited no response, although this did not surprise me. It was characteristic of her to listen to me babble on and on. When I turned around to check on her, those brown eyes just looked at me with devoted kinship. What a listener! While I had her undivided attention, I thought it best to get the hardcore realistic factors out of the way. "You realize that I am creeping up there in years and it is going to be quite an adjustment going from sea level to 19,340 feet in four days after flying half way around the globe." Again, there was no response other than sheer attentiveness, so I continued. "I have skied on many occasions above 12,000 feet, but another 7,000 feet is a factor to be seriously considered, wouldn't you think?"

"The extreme temperature is another serious consideration. Are you aware, Friend, that I have never slept on the ground in a tent where the temperature was lower than the temperature in the freezer compartment of my refrigerator?" I did not elaborate by adding that I had only camped out once in my entire life.

I expected a reaction of some nature by this point, but got nothing, so I decided to revisit the elevation issue. "I understand that the altitude sickness can be a bear. Did you know that areas within your brain and lungs swell from fluid that leaks out of your capillaries? It is because the partial pressure of the atmospheric air is not great enough to keep it within the vessels." I sensed that I had lost Belle on this one, although I've learned to never underestimate what someone else may comprehend.

At this point I had covered the major reasons for not making the trip. I must admit that verbalizing the lists was sobering, as the spoken words seemed to make it clear that my answer to the invitation must surely be "no."

The waters below Lesner Bridge were presenting major problems with the stability of the kayak, so it was time for a break in the conversation. I needed to stop talking and start concentrating on keeping the two of us from being tossed overboard. Once we cleared the bridge, I decided that perhaps an intermission was in order so I found a good place and paddled ashore. We both were looking forward to stretching our legs. As always, she bounded out of the kayak before we got to the beach. There is something instinctive about standard poodles that tells me she loves the water as much as I do. She became promptly engrossed in her swimming and I knew then that she was taking the opportunity for a mental break from our conversation. I left her alone to run, swim, and frolic in the sand. She was a tremendous "gift" and surely the answer to my prayer.

A Trek to the Top of Mount Kilimanjaro

"It's time to go," I finally said. My buddy came running over and we boarded the kayak. We eased the kayak back into the bay and resumed our rhythm. I was anxious to move to the next topic, which encompassed the list of reasons "Why I Must Make the Trek." I was ready to think beyond the barriers and consider the compelling reasons why I should attempt to climb Mount Kilimanjaro.

I began by saying, "A trip of this nature would add to my scientific knowledge. It would encourage me to read and study topics such as altitude sickness, oxygen deprivation and biome diversity. July will be an ideal time to climb, as it will be winter in Tanzania and I can escape the oppressive heat and humidity of Virginia Beach. It's a non-technical trek, which makes it even greater for me." I added, "It will be an opportunity to see if my exercising and health-conscience attitude sustains me in a harsh environment." That was essentially the end of the list of reasons why I should accept the invitation to climb Mount Kilimanjaro. In comparison, the "pros" were so few and seemed paltry compared to the extensive list of "cons." There was a lull in the conversation, as if my friend expected me to add substantially to the compelling evidence of why I should go on the trip. I had no further data to offer.

My true reason for desiring the experience was virtually impossible to record on a list or to share with anyone. How does one state that they need to reach out and possibly experience the essence of being alive. This time would be for me new sunrises to see, rare flowers to smell, and feelings to be felt which come only if I risk living my passions. The invitation spelled opportunity to apply my many loves: science, exercise, photography and meeting new people.

Nothing seemed to be eliciting a reaction from Miss Belle. I knew that surely we perceived things differently but this was turning out to be the exception, if silence represented consent. Along the shoreline, a tall handsome man playing Frisbee with his golden lab momentarily received her attention, but she returned to her phlegmatic demeanor. His perfectly sculpted body certainly received my attention.

Refocusing on the decision at hand, both of us knew that the scales weighed heavily against me accepting the invitation. However, both of us sensed that my answer would challenge the body of evidence. My buddy probably realized that my mind had been made up from the beginning, since my passions always seemed to overrule my willingness to find any idea "too expensive", "too difficult", or "too physically exhausting." Challenges seemed to moderate whenever I was passionate about accomplishing a goal or objective. She probably recognized my inner drive and desire to accept the challenge of climbing Mount Kilimanjaro. Belle had been with me on other occasions when I would sort out the pros and cons like the textbook prescribed, only to thank her for listening and tell her that I'd miss her while I was gone. She looked at me with her head slightly tilted as standard poodles often do which I interpreted as, "I'll miss you while you are gone and good luck!"

My friend had given her response and I could imagine that by now she was hoping for a miracle in the form of me changing the subject. As for me, I considered our trip a great success. I had talked and processed, and I had been given the insight needed to make this crucial decision.

As we glided through the peaceful waters of Long Creek, I knew that from this moment nothing about my life would be the same. I knew that the end of this kayaking trip marked the beginning of a far greater adventure. Climbing Mount Kilimanjaro would be a test of the mind, body, and soul, but my decision was made. Against the odds, I accepted the invitation.

ANN AND BELLE ON LONG CREEK

The Decision Is Made!!!

3

First Meeting

June 10, 1998

... I met with the other members of the climb for the first time today. The five of us sat down and exchanged our personal expectations. My impressions are positive. I feel that I made the correct decision to come aboard. We range in age from twenty four to sixty three. Four of us are educators and one is in medicine. Each person appears to be highly self-motivated and success oriented.

Our leader Harry, a tall, extremely sensitive high school AP government teacher has skillfully organized the trip. He has planned for the five of us to climb Mount Kilimanjaro and then to join another group of educators from Virginia Beach, Virginia in Johannesburg to travel to Botswana for a three week safari.

Harry, at forty-seven will be the solo male. He appears extremely physically fit. I find him brilliant, kind, and well traveled. He has recently met the love of his life, so he walks around with a perpetual smile on his handsome face.

A Trek to the Top of Mount Kilimanjaro

Donna, our youngest member works as a school psychologist. She is an effervescent individual with a great sense of humor. Her smile is contagious and it is easy to perceive that she will be an asset to the team.

The next member is Barbara, a nurse practitioner. Her manner of speaking has a special quality of compassion and concern. Her laugh is great; I look forward to hearing it. She plans to make the climb and return home rather than continuing on for the safari. She is forty-four. "Photography is one of my favorite things and I'm counting on capturing as much of the flora and fauna on film as possible," she emphasized.

I am the next oldest climber at fifty-eight; I have recently retired as a public school assistant principal. I now teach as an adjunct Biology Instructor at Tidewater Community College. I am

CAROL, DONNA, HARRY, ANN, BARBARA *Photo by Ofuru*

shy, sometimes quiet, and obsessive about almost everything which I undertake to accomplish.

Carol, the matriarch of the group is the only member of the team that I had met prior to today. She is a fellow public school assistant principal whom I have admired from a distance since her brave soul fought a courageous battle with cancer. She loves Africa. This will make her third trip to the continent.

I suggested that while we were getting acquainted with each other that the five of us might find it beneficial to meet on the weekends to hike and/or jog in Seashore State Park. Harry excused himself from the idea; he preferred to exercise alone. Donna, too, had a set routine at her gym and said thanks, but no thanks.

That left three of us. We agreed to embark on an opportunity to bond.

4

The Agenda for Travel & Trek

Day 1: Norfolk/New York/Johannesburg

Day 2: Land in Johannesburg, South Africa and spend the night

Day 3: Fly to Nairobi, Kenya: lunch, take van to Arusha, Tanzania, spend the night at the hotel in Arusha

Day 4: Leave hotel to meet outfitters for a briefing; travel to Machame Trail Head @ 6,642'; trek through the rainforest to Machame camp (9,774')

Day 5: Leave Machame – trek to Shira Camp (12,595')

Day 6: Leave Shira – trek to Barranco Camp (12,956')

Day 7: Leave Barranco – trek to Burafu Hut @ 14,924'. Rest from the time of arrival; leave at midnight for Uhuru Peak (19,340')

Day 8: Reach Stella Point (19,000'), to Elveda Point (19,270'), to Uhuru Peak (19,340'), descend Mweka Route to camp at Mweka (10,200')

Day 9: Hike down to Main Gate, lunch and take the van to Arusha

Back to Arusha for night

Day 10: Leave for Nairobi and on to Johannesburg, South Africa

Travel decisions are made, dreams are dreamed and the hope of fulfilling each individual goal lies ahead. We wish each other health and good luck. It is all UP HILL FROM HERE!

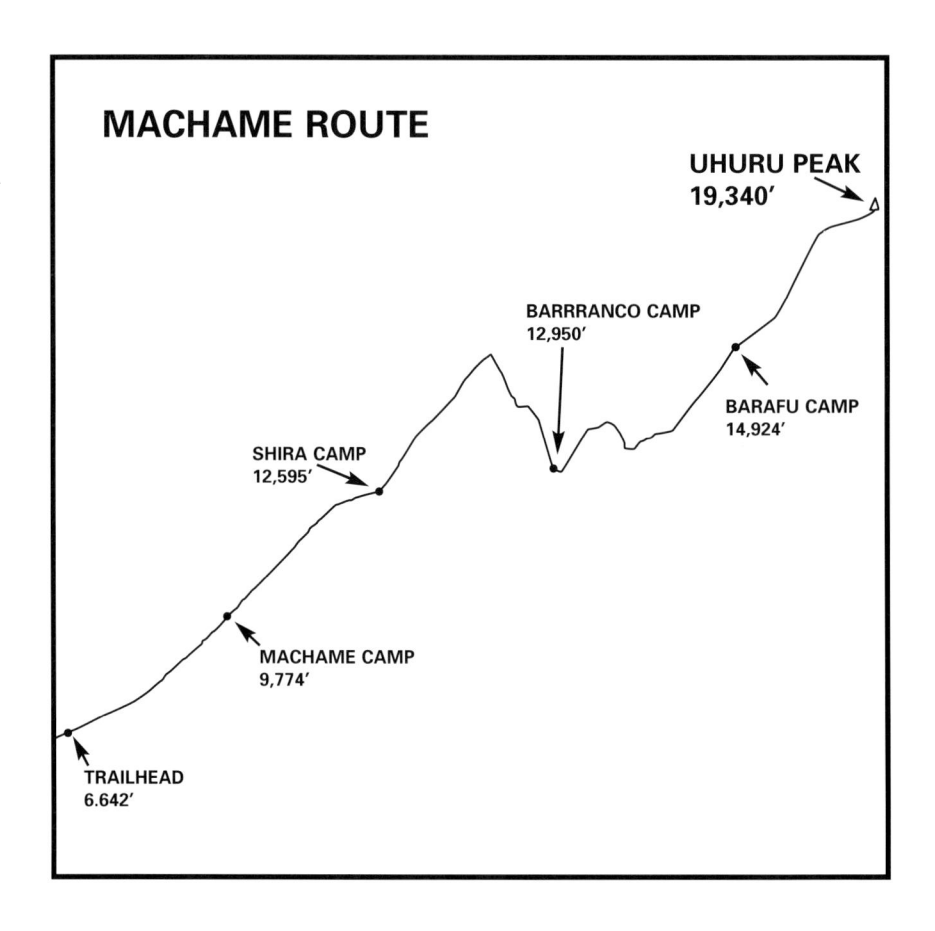

MACHAME ROUTE

UHURU PEAK
19,340'

BARRRANCO CAMP
12,950'

BARAFU CAMP
14,924'

SHIRA CAMP
12,595'

MACHAME CAMP
9,774'

TRAILHEAD
6.642'

5

Acclimation Agenda

It appears that the acclimation period will be the day we hike from Shira to Barranco. The logic is convincing, but it worries me. An altitude increase of only 400' for the purpose of acclimating would be "a sound bit of reasoning" if one resided in the Swiss Alps. I realize that this is not a realistic goal for this trek.

Dr. Bezruchka in his book, *Altitude Illness: Prevention and Treatment*, states, "A common strategy is to climb high during the day but descend to sleep at an altitude not more than a thousand feet above where you slept the night before."

The reality of going from zero to nineteen thousand feet in a matter of five days concerns me, but I know the agenda has been set. It is out of my control.

He adds, "For every three days above 10,000 feet, add an extra day at the previous night's sleeping altitude to the schedule. Arrange the itinerary to acclimatize slowly."

6

From Here to There

For me time was dragging at the pace of molasses on a cold North Carolina morning. I would prepare for my classes at the college, teach them, and rush home to devour another article about the climb. The next day, I would exercise, rest, and read more articles. My fun came in the form of mental exercises that I would play with maps that I had extracted from the web. I learned the characteristics of each biome that I was to trek. This included any plants, trees, animals, or unique climate changes. My mind recorded facts and figures about the great mountain. I purchased a beautiful red leather dairy with shiny pages in which I would record every fact about the trek.

I decided that it would be fun to compare and/or contrast my sea level origin "here" to the 19, 000 feet above sea level destination, "there." To understand where you are going, you must first know from whence you have come, so I have read. Doing additional

A Trek to the Top of Mount Kilimanjaro

research would constructively pass the time.

My home at sea level is located a few miles from the Atlantic Ocean. It is a magnet in the summer that attracts millions of vacationers. Visitors who, unlike myself, enjoy sitting out in the hot air, soaking up the UV and IR rays of the sun, and returning to their homes with plans of coming back again next summer.

Virginia Beach, Virginia is the greatest spot on the planet for eating blue crabs, catching fish, and watching dolphins swim along the shoreline in the mornings and evenings. Almost any activity is available to the sports-minded individual. Jogging, sailing, golfing, scuba diving and kayaking are among only a few of the accessible activities. It is a young child's summer paradise. They can spend endless hours digging in the crystalline sand, making castles, and watching the tide wash their creations away.

ANN & BELLE IN VIRGINIA BEACH

A Trek to the Top of Mount Kilimanjaro

The miles of seashore that are a few hundred feet from my front door have been the benefactors of four, possibly more, ice ages. The bay area was home to the early Native Americans. The ocean's shoreline, whose history spans back millions of years, has been altered by nature but remains an unending historical record for scientists to study.

Mastodon bones have been recovered near the oceanfront which have been carbon dated to be 19,000 years old. Inland, the great Rice's Fossil Pit in Hampton, Virginia has been a wealth of data for young school children and for scientists. Its history dates to the Miocene Era. The fossil pit houses whale bones, scallops and clam shells, and numerous other reminders of the past that help document the existence of seas freezing and thawing in accordance with glacial rising and falling.

With all of this unique, wonderful ancient history, it is, however, at sea level, with the exception of a giant man made garbage mount we have named Mount Trashmore. There is an escarpment ("scarp") left by a glacier in the nearby city of Suffolk, Virginia. The scarp which is 25 feet above sea level is considered to be a gentle climb.

Oxygen is plentiful and altitude sickness is unheard of. On a typical summer day, you'll get sunburned. On a windy winter's night, you'll put on a warm coat and have a great excuse for turning on the gas logs. Otherwise, our weather is rather predictable, except for the potential hurricanes that flirt with our shorelines.

Nor'easters that result in tidal flooding are the largest threat to our beaches. Our seasons fold gently into each other and extremes are rarely noted.

A Trek to the Top of Mount Kilimanjaro

In contrast to the sand dunes of First landing State Park (formerly Seashore State Park) where I jog and bike, Mount Kilimanjaro National Park is four miles wide and four miles up. It is the greatest free standing mountain in the world. Mount Kilimanjaro's highest peak is Uhuru Peak at 19, 344' on Kibo crater. It is one of the few summits of the world that can be reached easily by trekkers with the right physical preparation. The park's second highest crater peak is Mawenzi (16,894'). Unlike Kibo, it must be climbed only by skilled mountaineers.

Kilimanjaro's formation is related to the formation of the Great Rift Valley. It dates back to the early Pleistocene Era, approximately one and a half million years ago. The mountain's volcanic activity was centered in three areas: the craters of Shira, Kibo, and Mawenzi. The summits of each are around 16, 400' or more. Shira first, then Mewenzi became extinct. Kibo only has remained active.

Kibo's last eruption occurred 100,000 years ago, when it was estimated to reach a height of 19,352'. During this eruption the lava covered parts of the Shira crater, making the plain which is called the "Shira Plateau." The eruption also created the flat-like lava plain called "The Saddle" which extends toward Mawenzi. It is reported that at the present, Kibo is still a dormant active volcano with a strong sulfur smell that often is noted near the crater.

There are numerous accounts for the origin of Mount Kilimanjaro's name. The word "kilima" means "small hill" and "njaro" means "greatness" in Swahili. In Chagga, the language of the people who have settled around the lower slopes, "njaro" means "caravan."

Although the African people have occupied the lower slopes of the mountain for thousands of years, few written references were made to Kilimanjaro until the 19th century by Ptolemy, an Alexandrian geographer and astronomer. The missionary Johann Rebmann described the snow-capped mountain in 1849. Professor Hans Meyers was the first European in 1889 to ascend its peak.

Areas below 6000' have been cleared for cultivation, but the forest at around 8,800' remains up to the National Park boundary. The vegetation turns to *moorland* which is dominated by giant heather and tussock grasses at about 13,800'.

After the moorland, the land gives way to the bare *Alpine desert* beyond 16, 000'. It is here that the ground is devoid of any growth with the exception of lichen.

Mt. Kilimanjaro has a full range of climatic conditions which accounts for the unique plants and animals that inhabit its slopes. At around 13, 000' the temperature can be over 15°C during the day and freezing (0 °C) at night. Summit temperatures range from 5°C in the day to as low as -22°C at night. During the day, the cloud level rises from 7, 000' in the early morning to 16,500', and falls back in the evenings. Above 16,000' the mountain often remains clear all day with high wind speed. Kibo is usually above the clouds whereas Mewenzin is in them.

As reviewed earlier, and in summary, the routes to the summit are varied and all require a porter or guide. The most popular route is the *Marangu* Tourist Route. The *Mweka* Route is the fastest, steepest, and most direct walking route to the summit. The *Umbwe* Route is relatively short, steep, and scenic. It is not considered an easy ascent. The *Machame* Route is the most scenic route up the mountain. The *Shira* Plateau Route has its advantage since during the dry season the Shira Hut can be reached by four-wheeled rovers. Acclimation is a problem with this route and additional time must be factored into the trek following the vehicle ride.

The plants to note start at 5,248-9,840' which is the wet side of the mountain. This less steep slope contains giant camphor, fig trees, and podo. The steep side of the river valley has palms, ferns, lianas, and epiphytes. The dryer, western northern slopes (6,560-9,184') have cedar, podo, and wild olive.

At 9,840' the forest will be gone and the giant heather (a belt of tussock type grass) below the *moorland* proper will be apparent. Higher up there will be more giant heather festooned with old man's beard. The real *moorlands* (13,776-15,744') will have small tufts of grass and patches of lichens. Since true *Alpine Desert* has little rainfall, there is little growth to note there.

Certainly there is a vast contrast between Virginia and Tanzania, but both proudly have one thing in common. They are the home of many historical first beginnings. Early man first appeared about 10,000 years ago on the mountain in Africa and in Virginia. Africa, with its cradle for mankind, and Virginia, home to American Indians. Bones and other fossils provide material for science oriented students. They are unique and beautiful and both have ushered in millions of years of evolution.

MASTODON TOOTH

Part Two

The Journey

"Rock and ice, sky and clouds. The more-than-mile-wide volcanic caldera lies below; the smaller Reusch crater rises in the center like a portal to the underworld."
– Mel White, **National Geographic Traveler**

7

The Big Bird

Carol and I left Norfolk for New York. We arrived at JFK to learn that our bags were in one place and we were in another. It was a difficult five hours! We spent "touch and go" moments trying to explain to the lady at the airport that we had to arrive with our bags in South Africa. She kept trying to tell us (in Spanish) that they would eventually arrive. It was frustrating because without the contents of the bags, there would be no climb.

We took a taxi to the Holiday Inn. When we arrived, the lady at the desk said that she had no record of a reservation. Things were going great for us. I had the confirmation letter so we got over this bridge. The airlines located our bags that evening so there was another problem dissolved. We left the hotel for the airport and boarded our flight to Johannesburg, South Africa. We sat in the top bubble area of an African Air 747. I sat beside a man from Utah who was going to a private game reserve to kill an

African animal. "It doesn't matter which one," he said. He explained that to fire a single bullet would cost him $25,000. In the eight seats in front of us sat was the rest of the "killing party." They had boarded with long silver cases containing their equipment. I found the situation disheartening.

We landed in Nairobi, met Harry, and went to a nearby spot for lunch. Shortly after, we were met by the van that drove us to Arusha.

This was the wildest ride of my life. We would go on and off the road depending upon what was crossing it. A group of giraffes ran in front of the van, and instead of slowing down, the driver veered into the ditch and then back onto the "road."

We saw beautiful individuals, dressed in their native clothes, tending to goats in the fields. We saw children in the school yard dressed in their shorts and shirts playing a game of soccer. Occasionally the van would stop at villages along the side of the road and the native women would come out to sell us their exquisite handmade jewelry. Instead of money they wanted to barter for pens, aspirin, or t-shirts.

8

First Glimpse

A short time before landing, the pilot announced that we would be flying over Mount Kilimanjaro. I was so excited I didn't know what to do. The four of us jumped up from our seats and tried to catch a glimpse of the crater. It was unbelievable! I had seen this very image in a black and white picture in Hemingway's,

MT. KILIMANJARO FROM THE AIRPLANE *Photo by Carol Anthony*

A Trek to the Top of Mount Kilimanjaro

Snows of Kilimanjaro, a few weeks earlier.

Upon arriving in Arusha, we checked into the Impala Hotel and showered for the last time for seven days. The next morning we were collected by the Land Rover and taken to the outfitter for an orientation. From there we were driven to the Machame Gate.

9

The Rain Forest at 9,450'

The ruts in the road leading to the Machame Gate were deeper than I had ever experienced over any terrain in the United States. The chances of us reaching the Machame Gate felt slim to none. But from out of nowhere came small children and young men who pushed the Rover in and out of the ruts over three miles. We reached the gate without turning over, which was the first of many miracles.

It was at the Machame Gate that I met our Head Guide, Francis. He smelled strongly of alcohol, smoked short little African cigarettes one right after the other, and was the most charming, beautiful human being that I'd ever met. He named each of us as we stepped from the Rover. I was "Mama" and he said it affectionately and with sincerity.

I thought that we would never leave the gate. There were tons of paper work to be completed and time seemed to have no importance to anyone except us crazy, out-of-our-element, Americans. During our wait, I snacked on a Power Bar and put my camera, luckily, into a Zip-Lock bag and tucked it in my backpack. Francis came into the little room where we had been waiting for an hour or two and gave us a pep talk and the only piece of advice that he offered during the entire climb, "You must go slow, 'pole, pole,' " he said. He offered us a map of the mountain, and told us to "saddle up" for we were off and climbing.

The outfitter had assigned us two guides. The other guide, Ofuru, had ridden over in the Rover with us from Arusha. I had noticed that he spoke limited English. He was a young handsome man with a lovely smile. He wore a pair of Levis and Converse tennis shoes. He told us with a huge smile on his face that he was married and had two young girls. One was Emily and the other Ester. His demeanor was beautiful and his kind, intelligent features were accented when he spoke.

When we started the climb, he put on a light wine colored wind jacket. By the end of the climb to Machame Hut, I had renamed him Angel, which he seemed to like. It became immediately obvious to me that he was my angel, and his influence would inspire me throughout the trek.

This sea level adventurer grasped at once the value of placing one foot down and then taking a long slow breath. I learned in the first hour of the first day that patience and perseverance would be my only salvation. And when Francis, our guide, said, "pole, pole," I would do just that (slowly, slowly). I was like a fresh water fish that had been placed in the ocean. I knew that old habits would have to be abandoned. My mind could not take a holiday

A Trek to the Top of Mount Kilimanjaro

and wander off as it does while I stroll in the state park near my home. I would have to discipline my every move for signs of dehydration and fatigue. I must think if I were to survive this journey.

The rain forest was all that I had read that it would be – rain and more rain. Mud and roots that would take you right in were everywhere. It was a good thing that I had seen pictures of its trees and flowers, because I never looked up from the start to finish of that day's trek. I kept thinking about how much fun it would be if I could have looked up and/or around. We were on the wet side of the mountain where enormous giant camphors and fig trees grow. Palms, ferns, lianas and epiphytes grow on the steep, vertical side of the river valleys. I overruled my desire to take in the beauty around me and kept my eyes on the path in front of me. Overhead I could hear the Harlaubs Turaco calling. Crowned hawk, black duck and green ibis are sometimes seen.

There were challenges for me after the first five minutes of the

Photo by Ofuru

A Trek to the Top of Mount Kilimanjaro

climb. I could breathe all right, but I had packed too many pounds in my backpack to manage the balancing act. Ofuru asked if he could take my knapsack after hour five of my scramble through the forest. This did nothing for my ego, but it was my lesson for the day. The only thing that would go into my knapsack would be tissue, a roll of film, a power bar, and a bottle of H_2O. My camera would go in front, around my neck and under my shirt or jacket, for balance.

Ofuro and I reached the camp about an hour before sunset which meant the temperature had quickly dropped. Harry, Donna, and Barbara had arrived earlier, and Carol was still on the trail with Francis. Our tents had been assembled and our packs were scattered about in large yellow plastic bags. I recovered mine and dragged it to the tent.

The next hour was difficult for me. I was wet from perspiration and having a hard time thinking out the sequence of moves that I would have to take to get into dry clothes before total darkness set in. My backpack had filled with water but my camera was dry in the plastic bag.

All my papers (maps, notes, airline tickets, passport, money) were saturated. I manage to get re-dressed and place all my wet clothes in a garbage bag that I had brought along, but I could not locate my flashlight. I ran my hand slowly over the contents of my large bag several times before I realized that it was in my backpack, and the reason that I had not noticed it was due to the fact that it was in the water at the bottom of the bag. Luck was with me for it was a light that I'd used for underwater scuba diving, so it worked great when I twisted the top to turn it on.

A Trek to the Top of Mount Kilimanjaro

In the mean time, the cook delivered a plate of pasta, a pot of hot tea, and a little blue plastic cup. My sleeping bag was calling me, so I hopped in and waited for Carol and Francis to get to camp. It was raining harder now, but the sound was peaceful. Since I was dry and warm, the rain was no longer an enemy.

Seconds after I dozed off to sleep, Francis and my Carol came into camp. She got undressed, into her sleeping bag, and promptly fell off to sleep.

She was shaken over the ordeal. I learned the next morning that she and Francis had spent several hours standing in the dark waiting for Orufu to arrive with a flashlight. He was to have gone for anyone who had not arrived in camp by 6:00 p.m., giving them time to get there before sunset. But the instructions had been misinterpreted. Even in Chagga it was easy to note that Francis was highly displeased with his assistant.

10

Machame Hut 12,464'

The next day was cool and crisp! Breakfast was spread out on a colorful cloth on the ground just waiting for us to devour it. I honestly think that it was the tastiest meal that I had ever eaten. We took it down to the table cloth in record time.

It became quickly apparent that we were out of our element, or to be more exact, we had less of the element oxygen. At sea level, one breathes air that contains 20% molecular oxygen and 85% nitrogen. This morning the air equated to 50% of the 20% which tallies up to one-half of what I was accustomed to breathing. When I began to walk, I thought for a while that I was fighting a losing battle. But the routine of "one foot down and take a breath, and then another foot down and take a breath" started to work.

The terrain was simple, but extremely different from any that I'd ever seen. There were beautiful yellow flowers from *Senecio kiniensis*, *S. cottonii* and *Helichrysum cymosium* and red flowers from

A Trek to the Top of Mount Kilimanjaro

Gladiolus watersonioides and *Kniphofia thomsonii*. Large rock formations were teaming with lichens and heather. Tufts of long grass characterized the moorlands that we were crossing westward on our way to Shira.

Yesterday's lush forest with its muddy, root covered trail and today's moorland grasses were in stark contrast to one another.

Our dress was also different from yesterday's. Francis had on a great pair of Bermuda shorts and a baseball cap. Ofuru had shed his wine colored rain top for a light blue polyester jacket. The rest of us had on shorts. Harry had on his red gaiters and red jacket that he had gotten on a trip to the South Pole.

We stopped several times for water and Power Bars. We got our first look at Kilimanjaro's white top. I was so excited that I forgot that I was working hard at keeping it all together. It is amazing what you can do ...

GUIDES: OFURU AND FRANSIS

A Trek to the Top of Mount Kilimanjaro

MOORLANDS - HEATHER IN FOREGROUND

LICHENS ON ROCKS

A Trek to the Top of Mount Kilimanjaro

TUSSOCK GRASS CHARACTERISTIC OF MOORLANDS

ALPINE DESERT – ORUFU AND HARRY

11

Shira Hut 12,595'

Shira Hut was an interesting spot. After we arrived, in a matter of minutes, we transformed it into a Chinese laundry. All our wet clothes were draped over our tents, rocks, and signs. When the sun set, we had clothes that were a lot like little frozen pancakes. It was a challenge packing frozen clothes the next morning.

The spot provided me with many golden Kodak moments. The three sided outhouse, the open air breakfast/dinner tent with the little bird that waited for some bread, and the imposing magnitude of the tip of Mount Meru each reminded me of things and places that I had seen as a child.

We took turns laughing at each other's fingernails lined in a black background and the great hat hair. We laughed about everything; it was like we were a group of school kids out on the playground just living it up.

The night sky on Shira was overwhelming. The stars were so

A Trek to the Top of Mount Kilimanjaro

near. It was as if someone had rolled them down from their home in the sky for me to see or to touch up close and personal. I felt that I could reach up and select the constellation of my choice.

Along the ground were millions of Venice Ice Crystals. As I groped my way to the outhouse, at two in the morning, I could hear the crunch of ice under my feet. I took my flashlight and shown the light through the field of crystals that had arisen from the ground; there were trillions of prisms giving off their reds, yellows, greens, and blues.

We all stayed up taking everything in. The porters who camped in a cave-like spot in the side of the hill, played American music on their battery driven Sony Walkman (Michael Jackson was their favorite). I wanted it all to last forever. But eventually my body shut down and fell off to sleep.

MOUNT KILIMANJARO FROM SHIRA HUT

A Trek to the Top of Mount Kilimanjaro

The morning's breakfast was even more enjoyable than the day's before at Machame Hut. I was beginning, for a fleeting moment, to feel like an old pro at eating from the ground. At breakfast I recall drinking my oatmeal and feeling rather self-conscious over the fact, but after looking around, I noticed that everyone was drinking their porridge.

When we left Shira, I knew that I had my work cut out for me. We continued eastward toward Kibo, past a junction and continued east towards the Lava Tower. The trek was unlike the other two. It was like walking into another world. There were rocks, cliffs, magnificent trees, and breathtaking flowers.

To the north, during most of the day, we had the mountain to photograph. I went camera crazy each time Francis would stop to smoke. I took pictures of everything at every focal length. It was like being a child at a toy store the week before Christmas.

MT. MERU IN BACKGROUND

A Trek to the Top of Mount Kilimanjaro

To the south was the Great Rift Valley where human civilization had begun. I had never seen anything like it and had enough sense to know that I would never see anything like it again. I tried especially hard to record as many sites as my brain and camera would allow.

I could imagine the first human beings walking on the same path that my feet were touching. I visualized the human differences that hundred of years of evolution had produced. It was my favorite day for so many reasons.

We all laughed and shared tall tales about other trips that we had each taken. Francis was in such a good mood. He was a storehouse of knowledge which he enjoyed sharing with us. He had been leading folks up the mountain for 14 years. He told us about guiding the camera men and women from NOVA who made the video *A Journey to Mount Kilimanjaro*.

We passed through a bone chilling air pocket which was suddenly filled with tiny ice crystals. It happened so rapidly that I questioned whether I'd seen it or merely dreamed it. There was no way to take a picture of it because it all occurred so rapidly.

A personal moment came about mid-way into the trek that day. We were coming down a very steep bed of rock. I would classify it as an extreme slope if skiing. Fransis stopped to help Carol down the rock face and Ofuru ran ahead to help me. Fransis told him in Chagga to step aside, that he was helping "Mama."

As I came down at a rather rapid rate, Francis caught me. He hugged me like one embraces a lover that you'll know you'll never again see. It was a rare moment and one that I hope never to forget.

All of us stood there laughing and laughing. We were so happy

A Trek to the Top of Mount Kilimanjaro

and also so deprived of oxygen. These were once-in-a-lifetime moments that I knew that I would replay over and over again like songs that I love to hear. Fransis was feeling so proud of the good job that the elder ladies were doing on this most difficult day of trekking.

The weather was in our favor, but this would shortly start to change. We trekked what seemed to be an eternity. We were only going up slightly in altitude that day. Our path took us eastward to the Barranco Hut. It was located at 12,923' in a magnificent valley.

Since we were coming from Shira Hut which is 12,594', our slight increase in altitude would be for the purpose of acclimating. We were used to shorts, t-shirts, sandals and tons of air to breathe and the contrast was stark and often surreal.

I knew that this day would have little in comparison to the 4,000' climb that was expected of us on the last day from Barafu Hut. Since in reality, we were following almost the same topography line that entire day.

HIEM GLACIER ON WAY TO BARRANCO HUT

A Trek to the Top of Mount Kilimanjaro

I saw it all that day; Lava Tower or the Devil's Tooth and the Arrow Glacier. I remember Fransis would correct my pronunciation of Arrow. He would say it with style and a long "a" and insisted that I do the same.

He had climbed the Shiri Route with two Germans the week before and shared with us some great stories about their determination to do the trek in two days. I'm sure they made it if they were anything like the French couples that passed us each day like we were in reverse. They were using no hiking poles and sped by us on numerous occasions. We would hardly have the energy to say "hello," while they were fresh and appeared to be exerting no energy at all.

After climbing out of the valley, the trail ascended and ascended and ascended. At the end of the day, we all felt rather proud of ourselves. Carol and I had bounced back since Francis had taken control of the pace. I was hanging on because the pace was perfect for me. I would take a step and then a breath. I heard myself breathing and wondered if everybody else heard me. I finally said to myself "to hell with it"; breathing and walking slowly and methodically was working for me.

At the end of the day before descending to the Barranco Hut, I got the impression that Francis was lost. How wrong I was. He was pissed off. He did not like the fact that anyone moved ahead of him when they got the desire to do so. He sat down on a rock and smoked two cigarettes. That second cigarette said, "I am the boss!" I watched his eyes. He was a master at body language.

We finally started going again, crossed the top of the ridge, and bore northwest to the very welcoming Barranco Hut. No hut, no

A Trek to the Top of Mount Kilimanjaro

outdoor facility, just a tall rock, hot tea, and the greatest tasting popcorn that I had ever eaten in my life. I do recall that the "hut" was set in the valley and was a sheltered area below the spectacular cliffs of the Breach Wall.

SMOKE BREAK FOR FRANCIS

12

From Barranco to Barufu 14,924'

I'm sure this morning occurred, but I have no recollection of breakfast. I do recall that the moisture on the tents had frozen during the night, and for a moment, it looked as though it had snowed. I can't even remember what the dress of the day was but I suspect that we were digging into our warmer wear.

Francis had said, the night before, that we would be leaving earlier than usual,

**HARRY, BARBARA & ANN
BARRANCO BREAKFAST** *Photo by Ofuru*

A Trek to the Top of Mount Kilimanjaro

so be ready to get bouncing at 5:30 AM. He stressed that we had the hardest part of the climb to do on the first and the last leg of the day. The night went fast. And, as always, we were packed and ready to face the day as he indicated.

The morning air was crisp, fresh, and the sky was a "Carolina" blue. The stream that paralleled the trail was frozen on the top but veins of water flowed under the thin layer of ice, which appeared black. The crystal clear water was magnifying the black soil under it. It looked like the black oil that was made into a household term in the TV sitcom, the *X-Files*.

At noon we stopped for a bagged lunch and a restroom break. We were all chatting and laughing knowing that in a short while, things would change markedly. We finished our lunch and began what Francis called a short scramble to the top of the Great Barranco (13,120'). We traversed over scree and ridges to the Karanga Valley. We moved beneath the ice falls of the Heim, Kersten and Decken Glaciers. Everything was exciting. Eagles, possibly the Verreauz's eagle, flew above the cliffs. The air was bone-chilling one moment and warm the next.

Our pace was slow Francis was setting the pace for Carol and me. His feet were deliberately in sink with our ability to keep up. I could hear myself breathing which concerned me greatly.

As for the vegetation in this area only the hardiest of plants survive. The night's air dips to below 32° F and the temperatures in the daytime go as high 104° F in the direct sun. The *Spectrum Guide to TANZANIA* states, "The major problem plants must deal with at this altitude is 'solifluction'." Solifluction happens when the ground starts to freeze and the soilwater expands. This consequently results in the soil being moved upward and the plants have no choice but to go along for the ride. The few plants that do successfully live in this zone, such as the tussock grasses, keep their roots warm by utilizing the hair-like leaves which surround them. Other plants that survive in this zone retain their leaves and use them as insulation.

As we inched our way to the Barufu Hut, hours and hours passed. As strange as it might sound, I imagined I saw a wall of water to my right that was approximately ten feet high. It neither concerned me that the body of water was there or I perceived that I was scuba diving in it. I recall a floating sensation that was rather pleasant. I found myself vacillating between two activities, trekking and scuba diving.

Bezruchka, in his book *Altitude Illness, Prevention & Treatment*, identifies a situation of this nature as an "Altered Mental Status." After supper and tea, I recorded the incident in my notes, realizing that the occurrence was too bizarre to even mention to my fellow trekkers. Starr and Taggart state that, "A healthy person who grows up at sea level can make physiological and behavioral adaptations to a new environment that is markedly different from the one left behind by mechanisms of acclimatization." As I mentioned in Chapter 5, our acclimation period was obviously the day we hiked from Shira to Barranco (climbed high, slept low). We had not rested for an extra day or even a half of a day. I was, as my mother would say, "up the creek without the paddle."

I was experiencing Hypoxia, a state of oxygen deficiency in the body which is sufficient to cause an impairment of mental and physical functions. It's caused by the reduction in partial pressure of oxygen, inadequate oxygen transport, or the inability of the tissues to use oxygen.

Those of us who live at low elevations breathe air with oxygen being one molecule in five. When we travel above 8,000 feet or more above sea level, the Earth's gravitational pull is weaker, thus gas molecules spread out more and breathing becomes more difficult.

More specially, Hypoxia occurs when a reduction in the amount of oxygen passing into the blood takes place. It is caused by a reduction in oxygen pressure in the lungs, by a reduced gas exchange area, or exposure to high altitude. (Air Force Manual – http://www.webref.org/anthropology/h/high-altitude_mountains_sickness.h...). It is a form of high-altitude mountain sickness that can occur in persons not acclimatized to high altitudes. (http://www.wevref.org/anthropology/h/high-altitude_mountains_sickness.h...).

Climbing out of the Karanga Valley would take a toll on my body and soul. The four of us and Ufuru moved ahead of Carol and Francis at the end of the day. In retrospect, this was the lull before the storm; I was breaking the one rule that had been stressed from day one, "Pole, Pole."

We reached Barufu around 6:30 PM that evening. We had supper, rested and began again at midnight. We had 4,000' to go.

13

At Barufu Hut 14,924'

The Barufu Hut had a flavor entirely different from all the other spots we had camped. The dark side of the moon could not be as desolate. The hut was rather large. It was made of wood and the wood had darkened from the smoke from the cooking fires. It had a tomb-like silence about it. The porters were kicked back listening to music on headsets.

I remember seeing four tents all pitched on the edge of the cliff. It was probably the same cliff from which a lady, getting up in the middle of the night to go to the bathroom, fell to her death. The fourth tent was occupied by a young couple who were spending the night and the following day to rest.

Until Francis arrived in camp no one did anything. When the porters saw him, the place came to life. He screamed at everyone in Chagga. He did say in English the word "danger" which got my immediate attention. He was such a gentle man who most obviously was highly knowledgeable and respected. In all my journeys I'd never met anyone to compare to him. He was a small giant. his face was smooth and he sported a slight beard and mustache. I felt like I'd known him forever. Our spirits intertwined. I asked him if he had ever traveled, he said, "No, Madame, we have no recourses for that luxury." He had true class, and I will never forget him.

The tents that were relocated still presented a concern for me. I was at the "Oh, well," stage by then so I ate my supper and then lay in my sleeping bag until it was time for Francis to give us the agenda for the twelve-midnight ascent. I had researched this day for months.

WHAT'S AHEAD AT MIDNIGHT

A Trek to the Top of Mount Kilimanjaro

My glow in the dark watch said that it was 9:00 PM when Francis made his nightly rounds with our itinerary for the next day. We would be awakened in two hours, served tea and toast, and start our 4,000' trek up Barafu at midnight. He anticipated that we would be at the top by sunrise, sign the registry book which held the names of everyone that had reached the summit, and start down at once.

He added that we were to descend to Marango instead of Mweka, and he would inform the authorities of our plan change since new papers would have to be drawn up regarding the altered route. After leaving the top, we would go to Gillman's Point and from there take the trail to Kibo Hut. We were to be there by noon, eat lunch, and trek to Horombo Hut for the evening. After spending the night at Horombo, we would trek to the Park Gate for lunch and then on to the Impala Hotel for a bath and supper.

As I lay obsessing over the Barafu ascent, Carol told me that she was not going with us to the summit. I've always tried to respect the decisions of others and therefore I did not question why she had reached that decision. Her pace was my pace and I selfishly regretted that she was dropping off. I wanted her there because we had become friends over this journey and I realized what it meant to her.

Carol and I had an interesting conversation with the couple whose tent was next to ours at Barafu that evening. They shared with us that they were accustomed to high attitudes and were scheduled to rest and to take a short hike from their tent site tomorrow, and then climb to the summit the following day. We had arrived at Barafu around 6 PM and we were leaving around midnight. Carol had voiced her concern since day two about us having no acclimation time factored into the schedule. I told her the die was cast and we would have to live with our schedule.

I think before he finished talking through the details of our day/night, I may have zoned out for a short while. Things were beginning to get a little blurred for me; I could not place my whereabouts on the mental map that I had etched in my brain. I had followed every step of the way with the map in mind until yesterday. It is amazing what happens when your mind is deprived of its normal oxygen supply. Every inch of my body was suffering from fatigue. My face, hands and feet were swollen from *High Altitude Edema* (HACE). I felt that at any given moment that I would burst open.

We had not acclimated during any day of the climb, and it would take a miracle, or two, if we made it to our journey's end. I believe in miracles, and I would certainly be in need of one if we made it to the last leg of the trek.

When 11:30 PM arrived, I unzipped my last packeted bag of clothes and started to dress. It was to be the most important day for dressing properly since the temperature would range from 0° C to -22° C. This was the most severe environment that I had ever encountered. I would dress in layers. I started with cotton socks and over them thermal ski socks. Then capilene underwear, ski pants with a light fleece lining, ski turtle neck shirt, boots, gaiters and ski jacket designed for extreme skiing (No, I do not extreme ski), and a balaclava. I had never worn a balaclava before and wished later that I had settled for my fleece ski hat. With that out of the way, I could spend the remainder of my time obsessing

A Trek to the Top of Mount Kilimanjaro

about whether the battery in my head light would last at -22° C.

The idea of walking on snow at the equator was all that I had dreamed for over the past year. In a few hours I would be there and I would see it all. I would return to my Biology class and teach them first hand about biomes and oxygen deprivation. I would sit for hours and share with my mother details about the flowers, the birds, and the people. I'd be able to share with her what it was like seeing the sun rise from under the clouds. she was never too busy to listen to me explain where I'd been, what I'd seen and who I'd met.

14

What's Ahead At Midnight Barufu-4,000' Ascent

At midnight, the climb took on a different tone for me. When the physical pain had gone, the mental pain and confusion which comes at high altitude, set in. We all had our head lights on with our extra batteries tucked away just in case we needed them. It wasn't an hour before, one-by-one, our headlamps went out. Barbara had lithium batteries and hers held on the longest.

I got the impression that I was going to fry in all the clothes that I had on. My body temperature rose and I felt heat like I'd never before experienced. It was a great temptation to start unzipping everything and start taking layers off. I would unzip a zipper or two until the outside air cooled me down and then re-zip. My ski jacket had zippers everywhere, and I used each of them

A Trek to the Top of Mount Kilimanjaro

during that first hour of the climb. I was too afraid to do too much maneuvering of any kind, other than zip. I was scared that I would lose my pace or use up too much energy.

The mental game was in full gear now for I knew that the pace was quicker and we had to adhere to the time table. If we were late reaching the top the winds would possibly be too strong for us to summit.

I concentrated mainly on regulating my pace, but the stopping every five minutes to re-do the light situation was making it difficult. My other mental tool was to start reciting all the Bible verses that I learned as a child. My teacher had done a great job teaching me the verses. I kept repeating each of them over and over. The process helped me pass the time, but did little for the physical pain. "Though I walk though the valley of the shadow of death" took on a new and a very relevant meaning for me that morning.

DONNA & OFURU   **ACROSS REB/MANN GLACIER**

A Trek to the Top of Mount Kilimanjaro

Shortly more serious things than having a hot flash began to occur. Barbara stopped and indicated that she was getting hypothermia. Since she had an extensive medical background, we all took her statement extremely seriously. Francis, our head guide, sat her down and curled himself into the region of her stomach. The rest of us made a tent around her with our bodies. She had gotten hot earlier and had taken off her jacket leaving her long-sleeved fleece and her undershirt to keep her warm.

I said in a tentative tone which surely reflected my fear for her safety, "You know more about hypothermia and certainly more about yourself than anyone of us. Do you think we should go down?" I knew that so many things were running through her mind and heart. This was going to be a decision that must be reached immediately if she was to survive. She stated that her hands were freezing. I gave her my gloves. Francis made the decision.

Francis, being responsible for our lives, took her back to the Barufu Hut. Carol had chosen to stay back and was still at Barafu Hut. When they arrived, Francis instructed her to stand watch over her for one hour while she slept. At the end of the hour, he gave her some hot tea to drink, and then they started their descent to a hut that could render her medical help if it was still needed.

Donna offered me a pair of glove liners which I gladly took. I placed my last two hand warmers in the little silver gloves that she had loaned me. On my final bathroom stop, I had to sacrifice one of my two hand warmers for the sake of good hygiene. This decision and the little blue cup that I had borrowed from the cook, became topics for a good laugh for months to follow.

Our trip up Barafu continued with a few bathroom breaks

A Trek to the Top of Mount Kilimanjaro

SUNRISE – MAWENZI

along the way. Donna became nauseated; however the three of us kept moving all through the jet black night. Ofuru was now our head guide which raised my anxiety level on a scale of 10 to a 12.

At first light, which was like no other that I'd ever seen, the sun started inching its way through the ring of clouds that encircled "God's Mountain." The clouds were many hundreds of feet below us and glowed with a soft pale yellow tone. My hands felt frozen, but my brain was controlling my actions, so I stopped long enough to take a picture of the clouds below us. We saw a sketch of Mawenzi to the east. The site was awesome!

The ascent had its mysteries and its mishaps. The most fascinating puzzle that I encountered was the little red rocks. Little copper-red like stones lay in definite patterns on the edge of the glaciers. They were all similar in size and shape. I wanted to stop and pick one or two of them up but remembered thinking that if they had been placed in patterns for a purpose that removing even one would distort their possible meaning. I later concluded that they were placed near the edge of the glacier for a special reason to convey a massage to those who possessed the power to interpret their meaning. They may have been the recorded history of all that had occurred on the mountain top since the last eruption of the volcano. To this day I would love to go back to see if I could find them again.

The small red stones were possibly fragments of a meteorite that had arrived from space as pieces of an exploded planet. It could have been pieces of the Earth's long-age neighbor. Meteorites are rich in iron accounting for its reddish color (Van Rose and Ganeri, 1999).

The small dark grayish-black stone that I found later in the fold of my gater was possibly a Komatiite crystal. It was like nothing that I had observed on the mountain. Komatiite, I understand was a by product of a much hotter environment than modern lava. It was known to cool to form huge crystals which resembled blades of grass (Atlas, pg. 12).

Ofuru had us move on; we were running out of time. When the sun rises, so does the wind velocity. The top was near. It did not take a rocket scientist to figure out that our chances of reaching the summit were getting better and better. The sunrise had lifted my spirits. I was tooling along on an adrenaline rush. We were walking on the sheets of snow like professional mountain climbers who had ventured out for a 4,000' morning stroll.

A Trek to the Top of Mount Kilimanjaro

I looked up and there it was. But it was necessary to trek hundreds of yards more to reach the peak; Ofuru said it would take another hour and fifteen minutes. I was not sure if he understood the question, "How long will it take?"

We were down to one guide and the decision to continue would have to come from all of us. If one of us wanted to turn back all of us would have to consent. One member was extremely ill; she had been suffering from altitude sickness for hours. Another, I learned later, had become extremely cold. So we voted to start down.

The coin was tossed. I could see the summit, or at least I thought I saw it, and we were not going to make it. Donna's nausea (Acute Mountain Sickness) was becoming more and more severe. She was seriously ill, and we needed to turn around soon. If Francis had been with us instead of Ufuru, the decision to descend would have possibly been made much earlier than it was.

15

Time To Turn Around

I was tired and suffering from exhaustion, but I desperately wanted to finish what I had saved my last bit of energy for. The disappointment was overwhelming. The four of us turned and followed the path to H. Meyer Point, Stella Point, and around

DESCENDING TO KIBO HUT

A Trek to the Top of Mount Kilimanjaro

VIEW COMING DOWN

VIEW COMING DOWN

A Trek to the Top of Mount Kilimanjaro

Gillman's Point. We descended the Marangu Route to Kibo Hut. It was not a simple task moving down the mountain of scree. Usually defined as small loose stones, here the scree was ancient broken lava pieces. A tumble could cut terribly and we had no time or energy for another situation. We would go down a few hundred feet, stop and take a nap for about ten minutes, and then tackle another segment of the descent. We followed this procedure until we reached Kibo Hut at noon for lunch. We walked until a few minutes after 7:00 p.m. that day. Our tents were pitched at Horumbo when we arrived. All I remember was brief and to the point, "Get those smelly clothes off and drink this Coca Cola!"

The next day we hiked to the gate, said our goodbyes, and headed back to our hotel in Aurusa. We flew to Johannesburg the next morning to meet our friends from Virginia that had arrived for the safari.

16

Home at Sea Level

It was over in the blink of an eye. A friend and Belle greeted me at the airport in Norfolk, Virginia. We loaded the car and headed for my home overlooking the marshland. Of course, I insisted on stopping for fries and a peach milk shake. I ate the fries and Belle lapped the shake.

As I sat on my deck, the day appeared to be a typical summer day. The great blue heron was sitting on the osprey perch, the humidity was off the charts and the water in the bay was glistening in the hot sun's light. Nothing had changed here.

My heart was heavy with pain and disappointment, so I took the course of action that I take when I want to be alone with my hurt. I rolled the kayak down to Long Creek, put Belle's life jacket on her, and we took off for a paddle down the Creek to the Lynnhaven Bay. After about an hour, we went ashore on an isolated strip of sand. We were alone with the exception of several white

A Trek to the Top of Mount Kilimanjaro

egrets that didn't react to our sharing their spot. We swam and sat around like folks do on a hot, humid day. I had an apple, and Belle had a Milk Bone. I was home. A passing gull flirted with me for the apple core, and I finally gave in and tossed it up in the air. It was gone in a flash.

I wondered if Belle sensed my heartache. She was so tender and loving. Dogs make such great friends, and today she was giving it her all. She was my beautiful black standard poodle; my therapy dog who visited the sick and the lonely in residential homes. Today she was attending to her greatest fan's needs.

I have no idea how long I sat with my companion on the beach that day. I re-lived the last day of the trek over in my exhausted brain. We had not reached Uhuru Peak; we stopped at Elevada's Point which was a football field in length away from the summit, I calculated. We were down to one guide, so we all would go if one had to descend. We asked Ofuru to take our pictures and then we started the long four hour trek down to Kibo Hut for lunch. From Kibo we continued walking until we reached Horombos Hut where we spent our last night on the mountain.

It was the most physically painful day of my life. I looked down at my finger nails and they were a bluish-white bordered in a black ring. I took my finger tips and traced the blisters on my face and lips. All was not well. We had walked from midnight to noon to reach Kibo Hut, and we had seven more hours to go to reach Horombo Hut for the night.

I questioned my sanity for the first time since almost nothing in my body was registering pain to my brain cells. The point of total saturation had arrived. I had no choice but to keep my legs moving, knowing that sooner or later they would get me there and the rest of my body would be attached.

Belle was getting hot and it was time to complete the circle around the Lynnhaven and back up Long Creek before the mosquitoes started feasting on us. As we paralleled the shoreline, I realized that we had company. A dolphin swam within yards of the kayak. What a beautiful sight; I was home, now.

As we glided past the marina, with all its elegant boats and suntanned bodies, I heard a familiar voice yell "Welcome back." I was back from a mountain with its halo of clouds and its stark contrasting beauty. I was back to my Bay, with its clams, oysters, and periwinkles. I would continue my life where I had left it a month ago with my friend and Belle, both of whom I love so dearly.

We headed for shore at the end of my street. Belle jumped out and swam toward Maggie. I took a welcomed beer from her hand as I moved close to my docking area. For a few moments, however, I paddled back out from shore to collect my composure and indulge in one quick moment of solitude and peace.

Nothing about my life would ever be the same. I had returned with patches of memories that I would turn into words. Tall men in their splendid red robes that symbolized their tribe tending to a few goats stood in the fields. Women coming up to our van when we stopped, asking for pens in exchange for the baskets or jewelry they had made and school kids playing soccer in the school yard waving as we drove past. All of these sites would be forever etched in my memory. These things of beauty were mine now.

A Trek to the Top of Mount Kilimanjaro

I had met a proud, intelligent man named Francis who gently wept when I gave him the gift of a necklace with a raised image of a sailboat on the ocean on one side and a mountain with a snowy top on the flip side. His English was far better than mine and he understood exactly what I meant when I pointed to the sailboat and said, "I started here," and turned it over and said "and there is where I've just been - thanks to you." From here to there, oceans apart but joined in the communion with the natural worlds we both held dear.

Epilogue

"and there ahead ,..., as wide as all the world, great, high, and unbelievably white in the sun, was the square top of Kilimanjaro."
– Ernest Hemingway

It became obvious to me when our airplane dipped and I saw Kilimanjaro for the first time why the Masai and other tribes refer to its summit as "Ngaje Ngai" (the house of God). To me it was the most splendid natural feature that I had ever observed. I know that truly my soul had come back to its origin.

Never had I given the idea of flying half way around the world, hopping off an airplane, and two days later start climbing a mountain that was 19,000 feet higher than the sand dune that my home was sitting on. The possibility of reaching the summit of the tallest free standing mountain in the world was a given barring a major catastrophe. I had been drawn to the idea of reaching the top since the moment that I was invited to join the other four climbers for the trek.

Having been raised in a moderate, conservative Southern environment by a mother who stressed that hard work and preparation would reap me any harvest, I developed an "anything is possible" attitude, as long as it wasn't illegal and didn't harm others. Loving to exercise had been my salvation and I excelled at almost any form of athletics. The balance of elements that were required to climb this mountain I possessed. The possibility that I would succeed was always in my thoughts.

I loved the idea of a trek to the roof of Africa versus a technical climb to one of the base camps of some Tibetan mountain. I especially craved the idea of walking through a different biome each day, seeing and drawing little pencil sketches of the lobelia, the giant groundsel and the *Impatiens Kilimanjaro* which grows no where in the world except on Mt. Kilimanjaro. A storehouse of knowledge would be out there. Only because of some imperceptible occurrence or incident would I not reach Uhuru Peak.

Into my formula I factored in my continual exercise routine, clothing, medication, and all the required shots recommended by the Center for Disease Control. I expected few obstacles.

Hours were spent reading articles and outlining what I would see and in which biome I would note it. One of literally many articles was sent to me by my mom who had underlined it in several spots and asked me to respond. It was a short piece by Carolyn Haga, the assistant editor for *Traveler*. She wrote, "Our writer and photographer did not suffer any of the symptoms of acute mountain sickness (AMS), which can be fatal. Their group never ascended more than 2,500 feet in a day, and both took small amounts (less than the recommended dosage) of acetazolamide (Diamox). The article had a section on "Selecting a Route." It stated, "The six-day MACHAME ROUTE is a good alternative to the more expensive Shira Route." (*National Geographic Traveler*, July/August, 1998).

A Trek to the Top of Mount Kilimanjaro

I must admit, I read it and set it aside. We had received our agenda, and I should have given the article more consideration. This should have been my first clue that the Outfitters had not added an acclimation day to our agenda.

There were issues of more importance to me at the time. I purchased a pair of boots and trekking poles both of which I used everyday while hiking in First Landing State Park. Other equipment such as gaiters, ski jacket, balaclava, and gloves, I already owned. I never analyzed the larger issues such as Outfitters with their suggested time lines, and the ramifications related to the route that we had chosen. My boots had to be broken in and my exercise route had to be maintained to the letter of the law.

EPILOGY ONE

All five extremely able-bodied individuals failed to reach the summit of Mt. Kilimanjaro. This fact left me with the feeling that somewhere in the scheme of things certain variables had been over or underestimated.

We trekked the rainforest to Machame, the next morning we trekked to Shira, the following day we move on to Barranco, the next morning we trekked to Barafu. We arrived at Barafu around 5:00 PM. At 11:45 PM, the same day, we started to the summit. We hiked from mid-night to well into the morning before reaching Elveda Point. From there we trekked the entire day, with the exception of stopping at Kibo Hut for lunch. We reached the Horombo Hut around 7:00 PM and left for the gate the next morning. Acclimation was not the winning word for the week for the Outfitters that we trusted our well being.

A Trek to the Top of Mount Kilimanjaro

EPILOGY TWO

I had been home now for almost a day and a half, and each time I sat down to write I would think of something else to do. Coffee maybe, the maps, yes, I need the maps for my calculations are on them. A pencil, not a pen for writing. What if I remembered it differently after it's down on paper? You have guessed it, I was scared. What of, I am not sure, yet. It appeared that my fears should have gone, or at least subsided. I should have been thinking about reality which was starring me straight in the face; I was home at sea level. I needed to get on with my life.

It had been over three weeks since I left Mount Kilimanjaro. The obstacles that I faced during the climb, such as the mental fatigue, the fluctuations of body temperature, and the constant control of how much energy I could expend appeared to be minuscule.

The desire to wipe away the essence of everything that I had seen and done for the last year and a half was there for me. So if I didn't record the beautiful flowers, the sharp contrast of each biome, the stars, the ice crystals, the friendships that I made, surely I could conceal my pain. I must cover up the anxiety because no one understands what it's like to be "there" and not touch the top of Mount Kilimanjaro.

My eyes, I thought, could see it (Uhuru Peak) that morning. It was the length of a football field away. The vote was the right one. So I sit here with the task of rationalizing one moment plucked from four days of my life.

EPILOGY THREE

Many people have experienced the heartache of "not achieving" an obtainable goal. I am sure that was what my mother felt when she had to drop out of college because the depression had destroyed the financial stability of her family. However, lacking a formal degree didn't impair her ability to make and to share with others her monetary means.

Lacking a chance to stand at the top of Mt. Kilimanjaro, was equally as devastating to me. I was bitter and searched for someone or something to blame, as she probably did also. I never dreamed that I'd say it but the ability to move on to other endeavors, of equal importance, came with time, as surely as it did with her.

If there were any lessons to erudite; I hope that I've learned the vital ones. First, I don't blindly trust anyone who has my life in their hands.

Investigating timelines, asking questions, and leaving nothing to chance has become second nature to me. Secondly, I clarify things that are conveyed to me. If it makes someone defensive or they interpret you as confrontational, so be it. Clarifying someone's intentions is like making sure that there's water in the pool before you finalize the decision to dive into it.

Lastly, I've learned to be kind to myself; I had given it my all. So what, if I didn't finish, my soul was at peace on the mountain. What greater gift than that?

The End

References

Adventure Tours and MA.CO. LTD, P.O. Box 322, Zanzibar, Tanzania.

Bezruchka, Stephen. *Altitude Illness Prevention & Treatment*. Seattle, Washington: The Mountaineers, 1994.

Center for Disease Control and Prevention, www.edc.gov/traveleafrica.htm.

Else, David. *Trekking in East Africa*. Lonely Planet, 1998.

Embassy of Tanzania, 2139 R St. NW, Washington, DC 20008. Personal Correspondence.

Hemingway, Ernest. *The Snow of Kilimanjaro and Other Stories*, 1939.

Maresh, Carl. *Influence of Moderate Altitude Residence (2,200 meters) on the Acute Mountain Sickness and Exercise Response during Early Hypobaric Hypoxia (4,270 meters)*, 1981.

Ordinance Survey KILIMANJARO, Tourist Map of Africa's Highest Mountain, published by Ordnance Survey, Southampton SO9 4DH, Great Britain.

Pieg, Paseal and Verrechis, Nicole. *Lucy and Her Times*. New York: Henry Holt and Company, Inc. pgs. 30-44.

Salkeld, Audrey. *Kilimanjaro to the Roof of Africa*. Washington, DC: National Geographic Society, 2002.

Savage, Mark. *KILIMANJARO 1:50,000 Map and Guide*. P.O. Box 444827, Nairobi, Kenya.

Starr, Cecie and Taggart, Ralph. *Biology: The Unity and Division of Life*, 10th Edition. Thomson Brooks/Cole Wadsworth Group, 2004.

Tanzania Tourist Board, P.O. B 2348, Arusha, Tanzania: 57-3842. Personal Correspondence.

Tombazzi, Giovanni. *New Map of the KILIMANJARO NATIONAL PARK*, 1998. Revised and published in Cooperation with Hoope Adventure Tours and MA.CO. LTD, P.O. Box 322, Zanzibar, Tanzania.

Van Rose, Susanna and Ganeri, Anita. *THE BIG ATLAS OF THE EARTH & SEA*. ADK PUBLISHING BOOK, 1999.